Responsible Use: Biotech Tree Principles

A publication by the Institute of Forest Biotechnology

Principles Version 1.0.B

Responsible Use: Biotech Tree Principles

A publication by the Institute of Forest Biotechnology

Raleigh, North Carolina | USA | Published by the Institute of Forest Biotechnology

The Institute of Forest Biotechnology
140 Preston Executive Drive, Suite 100G | Cary, NC 27513 | USA

Contents

Preface

Thank you for your interest in the *Responsible Use: Biotech Tree Principles*.

People have always had, and will continue to have, an interdependence with forests. Given the reality of a growing world population, more productive, healthy, and sustainably managed forests are needed. We rely on the services forests provide, like cleaning water and slowing climate change by absorbing atmospheric carbon. We need sustainably managed trees to produce paper, packaging, homes, food, and renewable energy. We need to keep our forests healthy and productive to fulfill all these needs and to protect forested areas from decline.

These Principles are crucial because biotechnology is increasingly being used on trees and in forests. These Principles were developed in recognition that responsibly used forest biotechnology has the potential to benefit society, economies, and the environment.

Today there are invasive threats damaging our forests. We face a changing climate, deforestation, and illegal logging. Forest biotechnology can be a powerful tool against many of these threats. Scientists have already designed biotech trees that are resistant to disease and changing climates, growth rates that produce more wood fiber with fewer inputs on less land than conventional trees, and biometric tools to police illegally traded timber. Today there are over one million biotech poplar trees with the Bt gene that were established on commercial plantations in China in 2003. Genetic work on cacao trees is being explored to help the species that is susceptible to viruses in much of the world. Similarly, biotech papaya trees saved that industry in Hawaii from being destroyed by the Ring Spot virus.

Hundreds of researchers and organizations around the world have helped to pioneer these technologies in a responsible manner. Forest biotechnology is also being practiced in new ways, in new places, and by new researchers. But not every nation has a robust regulatory system or opportunities for interested public stakeholders to engage in issues important to them. We need the Responsible Use: Biotech Tree Principles to help guide long-term stewardship of biotech trees regardless of where they are developed or used. We need these Principles to foster a higher standard in biotech tree management, biotech forest stewardship, and ethical behavior.

Through an open dialogue that continues today, a broad spectrum of stakeholders, including university researchers, conservation and environmental groups, and industry leaders, created the Responsible Use: Biotech Tree Principles that are guided by these core beliefs:

- Biotech trees should benefit people, the environment, or both
- Risks and benefits of biotech trees must be assessed
- Transparency is vital and stakeholders must be engaged
- Social equity and indigenous rights are important and must be respected
- Biotech tree use must follow regulations in the country of their application

These Principles are unique because they are the first, and only, guidelines that include the entire biotech tree lifecycle from conception to final product. The Institute of Forest Biotechnology will continue to manage this initiative in a transparent way. Please visit the website dedicated to strengthening these principles at www.responsibleuse.org or contact us directly with your ideas.

Thank you,

Adam Costanza – President

Susan McCord – Executive Director

Introduction

Our Purpose

The Responsible Use: Biotech Tree Principles (referred to hereafter as 'Principles') were developed to help protect forests wherever biotech trees[1] are used. These Principles are the first of their kind and were developed through a transparent, multi-stakeholder mechanism, to achieve the following objectives:

- Establish a high level of performance for managing biotech trees that is recognized around the world.
- Create a simple and effective set of practices so users along the biotech tree value chain[2] know how to use the trees responsibly.
- Increase societal benefits when biotech trees are used by promoting interaction and education between foresters, biotechnologists, and other stakeholders.

Embodied throughout is an understanding that biotech trees and their products should create sustainable benefits. Benefits may be derived from the biotech tree, its products, or scientific insight gained through forest biotechnology research. The Practices give users tools to help them enhance the benefits of forest biotechnology, mitigate risks and maintain the integrity of a biotech tree's history as it moves along the value chain.

Process

The Institute of Forest Biotechnology (IFB) developed these Principles from a wide range of input from international experts in academia, environmental organizations, the forest products industry, and government agencies. A global team of experts formed the Implementation Committee that guided the development of the Principles while numerous stakeholders provided critical input throughout the process. In total there were five large stakeholder forums, and dozens of discussions with Forest Biotechnology Partners and individualized meetings with environmental organizations to craft these Principles. The goal to launch a set of stewardship Principles before biotech trees were widely available for use was a time-limiting factor. To best balance the immediate need for these Principles with the process of engaging a broad set of stakeholders, these Principles will be revised to ensure there is additional stakeholder input and that the Principles keep pace with the science, dialogue, and stewardship of forest biotechnology. These Principles will be reviewed every three years after an initial review in 2012. Procedures for revisions are in the Appendix. Additional information about the process of developing these Principles is available at www.responsibleuse.org/process.

Information on the individuals and organizations that contributed to the development of these Principles is in the Appendix: Committees and Contributors.

---

1   The Institute of Forest Biotechnology defines biotech trees as trees developed through genetic engineering or which contain discretely engineered DNA, and their offspring. This definition is intentionally inclusive of both the process (developed through genetic engineering) and the resulting tree (containing engineered DNA). Additional detail on this definition is available in the Appendix: Definitions.

2   A value chain is a set of linked activities. It is so called because each activity creates additional value along the chain. In this instance it refers to the various research, commercial, and physiological aspects of biotech trees.

Core Beliefs

These Principles are in recognition that responsibly used biotech trees have the potential to benefit society, economies, and the environment in ways that other trees cannot. Central to these Principles are core beliefs that:

- Biotech trees should benefit people, the environment, or both
- Risks and benefits of biotech trees must be assessed
- Transparency is vital and stakeholders must be engaged
- Social equity and indigenous rights are important and must be respected
- Biotech tree use must follow regulations in the country of their application

5 Truths

Academia, conservation groups, industry and all other stakeholders who developed these Principles agreed on five "truths," on which the Principles are based:

- Forests are important to people and animals
- Biotechnology is a powerful tool
- Biotech trees provide the potential for unique and diverse applications
- Biotech trees raise personal, environmental, and cultural questions
- Biotech trees are being used around the world with different levels of oversight

Use and Limitations of the Responsible Use: Biotech Tree Principles

- Use of these Principles is strictly voluntary.
- These Principles are not a certification system.
- These Principles only apply to 'biotech trees' that the Institute of Forest Biotechnology defines as trees that are developed through genetic engineering or which contain discretely engineered DNA, and their offspring. Therefore, clonally propagated or traditionally bred trees that do not contain genetically modified genes are not considered biotech trees by the IFB.
- This document is designed to stand alone or to be used as a complement to other programs or regulatory systems.
- These Principles do not take precedence over international, regional, local, or organizational regulations. These Principles are additive to such systems and users should be aware that there will likely be areas of overlap that are not explicitly detailed in this document.
- If a user[3] is already fulfilling the requirements of one of the Practices in this document through a different system, or when stricter regulatory requirements apply, then no additional effort is required other than documenting how the Practice is otherwise fulfilled.
- It is not necessary to be a Forest Biotechnology Partner to be "In Accordance" with these Principles.

---

3   'User' refers to a person or entity applying these Principles.

- The Institute of Forest Biotechnology is not able to certify or otherwise audit the efficacy of any person or organization using these Principles.
- Responsible Use: Biotech Tree Principles, Responsible Use, responsibleuse.org, Forest Biotechnology Partnership, and IFB are trademarks of the Institute of Forest Biotechnology. This document and all material at responsibleuse.org are copyright by the Institute of Forest Biotechnology. No part of these materials may be reproduced without the written permission of the Institute of Forest Biotechnology.
- Refer to www.responsibleuse.org for the most up-to-date version of these Principles and additional supporting material. Additions, corrections, case studies, and the Principle revision processes will be available at that website.

Components of the *Responsible Use: Biotech Tree Principles*

Value Chain, Sections, and Steps

Components of the Responsible Use: Biotech Tree Principles are grouped in three ways. The entire set of linked activities is the value chain. These components function together to help users and stakeholders work within a holistic framework to use trees responsibly. There are four sections within the value chain denoted by colors along the top row of the diagram below: Laws and Requirements, Product Development, Tree Growth, Tree Products. Steps are individual components numbered 1 through 7. Steps are designed to be additive to those before them. All steps after the first in the value chain of biotech trees build upon information from a previous step. For example, step 5, 'Obtaining a biotech tree,' cannot be accomplished until a prior user has approval to use biotech trees, which is step 4.

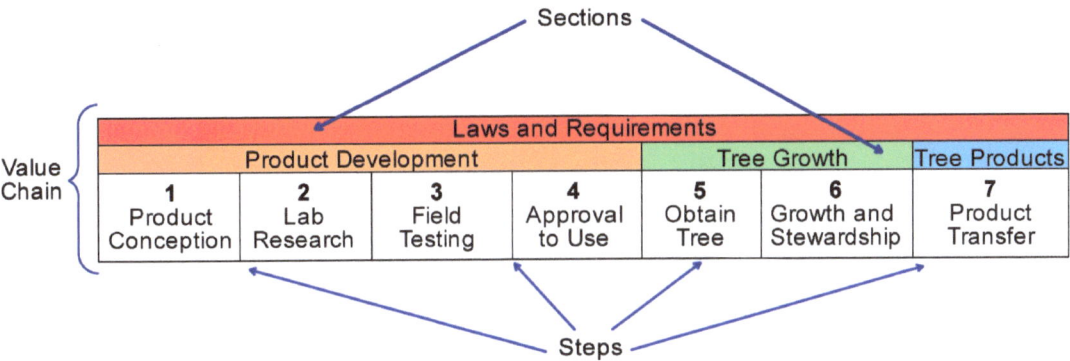

Practices – High Level

Practices describe what should be accomplished in broad terms. They are the performance measures for achieving stewardship. Each step has at least one Practice while some have more. Practices can be achieved in various ways, such as by following specific Actions that are described below.

Actions – Low Level

Actions detail how users could implement the Practices in specific ways. Most of the Actions consist of documenting results. The level of documentation would likely be proportional to how unique the biotech tree is. In general, it is useful to provide more documentation when a biotech tree is the first of its kind. In situations where detailed information is confidential and has to be restricted from outside parties, a secondary attestation by a party within the organization, a responsible party, or a peer, can usually be made available without divulging proprietary information. Some Actions reference alternative measures (designated as Alt in the margin) for widely recognized areas of overlap with established systems. If you believe an overlap and alternative measure should be included that is currently not, please submit it for consideration at www.responsibleuse.org/participate.

Recommendations and Discussion

Each of the seven steps has Recommendations that users can follow at their discretion. Users are encouraged to follow as many of the Recommendations as feasible. Each step also has a Discussion section that gives additional context and insight.

Tools

The Tools Appendix includes templates and worksheets to help biotech tree users complete Actions and Recommendations. These tools are intended to simplify implementing these Principles while maintaining a comprehensive and verifiable approach. The tools are generic while biotech tree uses are often unique, making it important to add information or modify a template or worksheet to suit a specific situation.

Additional tools, resources, and case studies will be developed over time at www.responsibleuse.org/resources. The IFB encourages ongoing discussion from users and sharing of best practices on the Responsible Use website. Please submit your comments at www.responsibleuse.org/participate.

In Accordance with Responsible Use: Biotech Tree Principles

Users can selectively apply any part of these Principles to their use of biotech trees, but those interested in achieving maximum effectiveness should complete each applicable Practice and associated Action. Once all Practices and Actions are completed, the user can publicly attest to being 'In Accordance' with the Responsible Use: Biotech Tree Principles (referred to as simply In Accordance throughout) for the respective value chain steps. Users should keep applicable documentation to verify their assertion of being 'In Accordance.' To promote confidence in the thorough application of these Principles, the IFB encourages all users to be In Accordance, to follow as many Recommendations as possible, and to make as much documentation relating to the application of these Principles readily available to stakeholders.

If users are members of the Forest Biotechnology Partnership[4], the IFB will assist them in implementing the Responsible Use: Biotech Tree Principles and publish any documentation they wish in support of an In Accordance assertion, and will keep a list of all In Accordance assertions from Forest Biotechnology Partners at www.responsibleuse.org/accordance.

---

4   More information about the Forest Biotechnology Partnership and how to become a member is available at: www.forestbiotech.org/partners.html

User Reference Sheet

This sheet is a quick overview of the Responsible Use: Biotech Tree Principles starting with the Core Beliefs at the top of the page, moving down through steps to complete, and ending with the optional attestation of performance at the bottom of the page.

Core Beliefs	• Biotech trees should benefit people, the environment, or both • Risks and benefits of biotech trees must be assessed • Transparency is vital and stakeholders must be engaged • Social equity and indigenous rights are important and must be respected • Biotech tree use must follow regulations in the country of their application
Value Chain The seven steps of biotech tree use – all of which are subject to applicable laws and requirements	

	All Users are Encouraged to	**Users that are In Accordance with the Principles**
Practices What should be accomplished for each value chain step	Address all applicable Practices	✅
Actions How the Practices can be accomplished	Follow all Actions for applicable Practices	✅
Recommendations Optional steps users are encouraged to take	Follow as many Recommendations as feasible	Optional
Tools Worksheets to help users complete Actions and Recommendations	Use and modify worksheets as necessary	✅
Attestation Alert the IFB of being In Accordance with these Principles	Publicly attest to completing all steps and being In Accordance	✅

Responsible Use Practices, Actions, Recommendations, and Discussion

There are seven steps in the biotech tree value chain. Each step is part of a section. The Laws and Requirements section is unique because it is overarching and inclusive of the entire value chain. Steps 1 through 4 are in the Product Development section, 5 and 6 are in the biotech Tree Growth section, and 7 is the only step in the Tree Products section.

Each step has at least one Practice, one Action, one Recommendation, and a Discussion. Practices describe what should be accomplished and are the performance measures for achieving stewardship in that step. Actions detail specific ways users could implement the Practices. To be In Accordance with the Responsible Use: Biotech Tree Principles, each Practice and Action must be completed for a given step. Recommendations are optional activities that further increase the level of stewardship, and Discussions provides insight and background for each step.

Laws and Requirements Section

Laws and Requirements						
Product Development				Tree Growth		Tree Products
1 Product Conception	2 Lab Research	3 Field Testing	4 Approval to Use	5 Obtain Tree	6 Growth and Stewardship	7 Product Transfer

Laws and Requirements categorically include all Responsible Use: Biotech Tree Principles

Every step in the value chain must follow all applicable laws and requirements first and foremost.

Practices

P0.1   Follow all applicable laws and requirements that apply to biotech tree use.

Actions

A0.1   Document the national, regional, and local laws as well as the institutional guidelines, and other mandatory requirements that apply to these biotech trees. Document adherence to the laws and requirements that apply to these biotech trees. Have the document recognized and signed by a responsible party[5].

---

5   A responsible party is someone who has both the technical skills to thoroughly understand what he or she is signing, and who has a vested interest in assuring that the information is correct. For example, a responsible party could be a colleague in the same field of work, or a supervisor in the organization, or a regulatory authority, among others.

Recommendations

R0.1   In countries where there are no laws, institutional guidelines, or other requirements that apply to biotech trees, use the Responsible Use: Biotech Tree Principles as a guide for stewardship of biotech trees while simultaneously working with organizations and agencies to implement stewardship measures for biotech trees at a national or institutional level.

Discussion

Adhering to applicable laws and regulations is common to almost every stewardship program. While it may seem obvious that any person or organization involved in forest biotechnology would automatically follow all applicable laws, it is useful to document what the laws are and how they are being followed. This step is especially useful for users in countries where there are no regulations governing biotech material. In such cases, documenting what steps were taken with justification will build confidence within the value chain that even in uncertain situations a thoughtful approach was applied through these Principles.

Product Development Section

Laws and Requirements						
Product Development				Tree Growth		Tree Products
1 Product Conception	**2** Lab Research	**3** Field Testing	**4** Approval to Use	**5** Obtain Tree	**6** Growth and Stewardship	**7** Product Transfer

1. Product Conception

The development of biotech trees to meet specific objectives.

Practices

P1.1   Develop biotech trees that benefit society.
P1.2   Evaluate the anticipated risks and benefits of developing these biotech trees.

Actions

A1.1   Document anticipated benefits from these biotech trees. The document should include the rationale for developing these trees and explain how the benefits are anticipated to outweigh the risks. Have the document recognized and signed by a responsible party.

A1.2   Document external, peer-reviewed information or regulatory information sources to estimate the risk and benefit of developing these biotech trees; a list of some resources is in Appendix: Risks and Benefits. Have the document recognized and signed by a responsible party.

Recommendations

R1.1   Initiate consultations with stakeholders including biotech tree regulating authorities.

R1.2   Broaden the intended benefits of these biotech trees to include the widest range of people and places.

R1.3   Review literature and databases that have information about the genes or gene constructs being considered for these biotech trees; a partial list of resources is in Appendix: Biosafety

R1.4   Consider how these biotech trees might be monitored to help mitigate potential risks.

Discussion

Since responsibly used biotech trees are developed to meet a specific objective, this step focuses on the intended benefits. Benefits may accrue through a variety of mechanisms including healthier forests, better forest products, enhanced carbon sequestration, and increased economic opportunities for landowners, among others. A thoughtful explanation of the intended social, environmental, or economic benefits will help create a clear product concept. There is tremendous value when research intent is aligned with social benefits and communicated both inside and outside an organization. Opportunities to engage broad groups of interested parties about the biotech trees are encouraged at this, the earliest step of the biotech tree value chain. These interactions are ideal situations for all parties to gain insight into how science could address specific social needs. Reviewing published information about the genes or constructs being considered may generate opportunities to improve on the benefits these trees may deliver. Such research could also uncover known risks that have to be considered and mitigated.

2. Lab Research

Test biotech trees in contained indoor environments such as research laboratories and contained greenhouses.

Practices

P2.1   Take steps to prevent the release of living biotech trees and associated living material[6].

Actions

A2.1   Document the steps taken to prevent the release of living biotech trees and associated living material including pollen, seeds, and agrobacterium. Have the document recognized and signed by a responsible party.

---

6   These materials are sometimes referred to as "Living Modified Organisms" or LMOs. LMOs are simply any living organism that possesses a novel combination of genetic material obtained through the use of modern biotechnology. See definition in Appendix for a more detailed description and references.

<u>Recommendations</u>

R2.1   Begin consultations with stakeholders, including biotech tree regulating authorities, or continue discussions already established.

R2.2   Follow appropriate national or institutional biological containment guidelines as described in Appendix: Biosafety

R2.3   Evaluate whether data being sought can be obtained from laboratory or greenhouse testing only.

<u>Discussion</u>

Users should establish a research environment that prevents the release of living biotech trees and their associated living material to the environment or other uncontrolled areas. Depending on the specific situation, it may be reasonable to expect that releasing these materials outside the laboratory will have negligible effect on the environment. However, to be In Accordance with this practice, take reasonable steps to prevent the release of these materials. It is also important to initiate stakeholder dialogues as early in the development of biotech trees as possible. Start or continue discussions with interested parties including authorities that regulate biotech trees.

3. Field Testing

Testing biotech trees outside of a contained indoor environment.

<u>Practices</u>

P3.1   Document the rationale for initiating field testing.

P3.2   Create a research plan for the field test that includes an environmental assessment and mitigation plan.

P3.3   Update the environmental assessment and mitigation plan for the duration of the field test.

P3.4   End the field test so that these biotech trees or their environmental impacts do not persist.

<u>Actions</u>

A3.1   Evaluate the tradeoffs of field testing these biotech trees versus continuing tests in a contained indoor environment. Document as many external, peer-reviewed sources of information as reasonably possible to estimate the risks and benefits of performing a field test. The document should include the rationale for performing a field test and explain how the benefits are anticipated to outweigh the risks. Have the document recognized and signed by a responsible party.

A3.2   Develop a research plan for the field test that includes the processes and procedures for: transporting these biotech trees to and from field tests, monitoring the field test, and the scientific tests to be performed. The plan should include strategies to contain gene flow based on the possibility of the genes establishing in the environment and the novelty of gene function. A long-term land use plan for the field test should also be included. Have the document recognized and signed by a responsible party.

A3.3   Evaluate monitoring data over the length of the field test. Document information that changes the research plan developed in Action 3.2 and any unanticipated changes on the site resulting from the field test.

A3.4   If relevant laws and regulations allow field tests to persist without devitalization, and allowing a field test to persist is in the research plan, document the ways in which the field test poses no significant risk to the environment. Otherwise, devitalize the trees in the field test by rendering them biologically inactive, following appropriate national or institutional biological containment guidelines as described in Appendix: Biosafety. Monitor the test site to ensure complete devitalization based on the biotech trees' physiology.

Recommendations

R3.1   Begin consultations with stakeholders, including biotech tree regulating authorities, or continue discussions already established.

R3.2   Design the monitoring plan commensurate with the novelty of gene function to verify the level of gene containment.

R3.3   Assess the risks and benefits of testing these biotech trees outside of a contained, indoor environment, and proceed with field testing only if the information required cannot be obtained in a laboratory or greenhouse.

R3.4   Promote stakeholder involvement in the field tests to engage and educate others in a real-world environment on the benefits and risks associated with these biotech trees.

R3.5   Monitor the test site for a minimum of at least one year and possibly longer based on the biotech trees' physiology, to ensure complete devitalization.

Discussion

There are more than one hundred field tests of biotech trees throughout the world, as field testing is often necessary in the biotech tree value chain. The reasons for performing research outside of a laboratory or greenhouse depend on many factors that the user should communicate to stakeholders whenever possible. In some instances, the purpose of the test may be to evaluate gene flow outside of the test area itself, but in most cases the objective is to see how the biotech trees will grow in a more natural environment. The intended use of the results of these investigations should be considered and documented whenever possible, because they will vary dramatically depending on whether the test is for scientific investigation and academic use only, or if it will be used specifically for biotech trees destined for commercial use. These practices should focus the researcher's attention on the need to design tests that achieve their intended purposes with respect to gene flow. Observing and documenting impacts on the site provides additional information on the effect these biotech trees have on the environment over time.

4. Approval to Use

When biotech tree developers have obtained authority to use trees in an unconfined environment.

Practices

P4.1   Enable a continuous chain of historic information about these biotech trees.

P4.2   Inform users about the biotech trees they are acquiring.

Actions

A4.1   Provide future users of these biotech trees with relevant information from application of the Responsible Use: Biotech Tree Principles by consolidating non-proprietary documentation into a single package.

A4.2   Provide future users of these biotech trees with a material profile of these biotech trees. Include information that typically accompanies trees when ownership is transferred. Include information that explains what makes these biotech trees materially different[7] from their non-biotech tree counterparts, and any unique information required to use these trees.

Recommendation

R4.1   Provide users with care and growing instructions, the intended use of these biotech trees, and any additional information that will assist subsequent users in applying the Responsible Use: Biotech Tree Principles.

Discussion

At this step the tree developer has received legal approval to sell or otherwise transfer and use a biotech tree in an unconfined environment. Following these Practices will help users of the biotech trees and their products know what they are purchasing and will give them the tools to be good stewards of these trees and their products. Forest Biotechnology Partners that are In Accordance can direct users to www.responsibleuse.org/accordance, where this information can be made available for them.

Tree Growth Section

Laws and Requirements						
Product Development				Tree Growth		Tree Products
1 Product Conception	2 Lab Research	3 Field Testing	4 Approval to Use	5 Obtain Tree	6 Growth and Stewardship	7 Product Transfer

5. Obtain Tree

Transferring living biotech trees to the next user.

---

7   Materially different and materially changed are terms used in this document to describe biotech trees or their products that are different enough from a non-biotech counterpart so that it warrants a distinction. The level of what is warranted is not strictly defined, but it should be based on stakeholder interactions and expert opinions.

Practices

P5.1   Ensure the biotech trees were legally produced and have proper documentation.

Actions

A5.1   Obtain from the original producer, seller, or entity otherwise providing these biotech trees, verification that these trees have received approval from the applicable regulatory authorities to be transferred. Obtain a material profile of these biotech trees that includes information on how to use them responsibly and explains what makes these biotech trees materially different from their non-biotech tree counterparts.

A5.1-Alt   Obtain biotech trees from a producer In Accordance with the Responsible Use: Biotech Tree Principles.

Recommendations

R5.1   Begin consultations with stakeholders, including biotech tree regulating authorities, or continue discussions already established.

R5.2   Obtain biotech trees from producers that are In Accordance with the Responsible Use: Biotech Tree Principles. A list of Forest Biotechnology Partners that are In Accordance is available at: www.responsibleuse.org/accordance.

R5.3   Those obtaining biotech trees should be aware of the trees' genetic modifications, such as changed genes or constructs.

Discussion

This step gives those obtaining biotech trees information to make informed decisions. It also provides a mechanism to ensure that the trees were legally produced and carry proper use and care information with them. Obtaining biotech trees from an organization that is In Accordance can reduce the time, cost, and documentation requirements of these transactions.

6. Growth and Stewardship

Planting, growing, and using biotech trees in the open environment.

Practices

P6.1   Use forest resources sustainably.

P6.2   Keep records of where these biotech trees are planted.

P6.3   Enable a continuous chain of information about these biotech trees and the land on which they are planted.

P6.4   If products[8] from these biotech trees are harvested and used, then follow the Principles in step 7.

---

8   Biotech tree 'products' are anything that is collected or harvested from a biotech tree, or the tree itself. See Appendix for examples.

Actions

A6.1   Follow applicable sustainable forest management practices. Document the management practices followed and the associated results. Have the document recognized and signed by a responsible party.

A6.1-Alt Obtain certification from an internationally recognized sustainable forest management system for the forest resources being used; examples of some are listed in Appendix: Sustainable Forestry.

A6.2   Develop and implement an oversight plan for these biotech trees at a spatial level appropriate to the area where they are planted. Include the spatial extent of these biotech trees, and inventorying techniques and frequency. Have the document recognized and signed by a responsible party.

A6.3   Provide documentation to new or contract landowners that includes the material from Practices 6.1 and 6.2 and information on what additional Responsible Use: Biotech Tree Principles were previously followed.

A6.4   Attest that products are not being harvested and used at this stage. If biotech tree products are being harvested or used then complete the Practices in step 7.

Recommendations

R6.1   Begin consultations with stakeholders, including biotech tree regulating authorities, or continue discussions already established.

R6.2   Follow an internationally recognized sustainable forest management standard; examples of some are listed in Appendix: Sustainable Forestry.

R6.3   Maintain as much oversight and information about these biotech trees as is feasible, such as stand location, planting dates, monitoring data, long-term oversight plans, and original supplier.

R6.4   Consider gathering additional information, or collaborating with other researchers to better understand the environmental effects of biotech trees. Examples of desirable types of information are available at www.responsibleuse.org/resources.

R6.5   Inform local tree farms and other landowners about these biotech trees.

Discussion

During this step biotech trees will have an impact on the environment; this is also true for non-biotech trees. The objective is to make sure the current user of the biotech trees has the tools necessary to make a positive impact by following established sustainable forest management practices. While some sustainable forest management systems explicitly disallow the use of biotech trees for certification, their guidelines on how to grow healthy trees and how to be good forest stewards can still be applied. Maintaining oversight is useful when transferring trees or land to future users who want to know their history.

Tree Products Section

Laws and Requirements						
Product Development				Tree Growth		Tree Products
1 Product Conception	**2** Lab Research	**3** Field Testing	**4** Approval to Use	**5** Obtain Tree	**6** Growth and Stewardship	**7** Product Transfer

7. Product Transfer
Transferring non-living biotech tree products.

Practices

P7.1   Users currently in possession of the biotech tree products should inform those acquiring the products about any material differences from similar, non-biotech tree products, and any additional product information they deem necessary.

P7.2   Users acquiring the biotech tree products should make themselves aware of any material differences from similar, non-biotech tree products.

Actions

A7.1   Document material differences of the biotech tree products from similar, non-biotech tree products. Provide additional use information if necessary. Have the document recognized and signed by a responsible party.

A7.2   Obtain documentation from the user in possession of the biotech tree products that details any material differences from similar, non-biotech tree products.

A7.2-Alt Purchase from a biotech tree user In Accordance with the Responsible Use: Biotech Tree Principles.

Recommendations

R7.1   Begin stakeholder consultations or continue discussions already established.

R7.2   Obtain biotech tree products from users In Accordance with the Responsible Use: Biotech Tree Principles. A list of Forest Biotechnology Partners that are In Accordance is available at: www.responsibleuse.org/accordance.

R7.3   Those involved in the transfer of biotech tree products should be aware of what makes the trees unique, such as changed genes or constructs.

Discussion

This step provides users acquiring biotech tree products with information to make informed decisions. In some instances, the product may be no different from those made from non-biotech trees. In other cases, the product may have unique aspects because its source was a biotech tree, in which case the customer should know what those characteristics are and how to properly use the product. However, only biotech tree products that are materially changed from their non-biotech tree counterparts require attention. Another option is to acquire the product from a biotech tree user In Accordance with the Responsible Use: Biotech Tree Principles to reduce the time, cost, and documentation requirements of such transactions. A list of Forest Biotechnology Partners that are In Accordance is available at: www.responsibleuse.org/accordance.

Appendix

Biosafety

Widely recognized biosafety guidelines that are applicable to biotech trees include, but are not limited to, the following:

- U.S. National Institute of Health Appendix P – online at: http://oba.od.nih.gov/oba/rac/guidelines_02/APPENDIX_P.htm
- UN Living Modified Organisms Article 6 Transit and Contained Use; Article 18 Handling, Transport, Packaging and Identification – online at: http://bch.cbd.int/protocol/text/
- Virginia Tech's Practical Guide to Containment, Greenhouse Research with Transgenic Plants and Microbes – online at: http://www.isb.vt.edu/Containment-guide.aspx

Industrial Biotech Tree Products

The Responsible Use: Biotech Tree Principles can be used as a starting point for biotech trees that produce industrial products or accumulate toxic materials. However, the scope of these principles does not fully cover the special stewardship requirements of these categories of biotech trees. Please refer to: www.responsibleuse.org/resources for the latest guidance available regarding these types of biotech trees.

Revisions

Refer to www.responsibleuse.org/process for additional information about the revision process for the Responsible Use: Biotech Tree Principles.

Timing: These Principles are dynamic and will be revised on a regular basis.

Scope: All aspects of the Responsible Use: Biotech Tree Principles are eligible to be revised, including the revision process itself.

Transparency: The IFB will announce when public comment periods are open and the overall progress of a revision online at www.responsibleuse.org/process. All material for public comment will be available online.

Risks and Benefits

Guidance on the risks and benefits of biotech trees is an ongoing effort for the Institute of Forest Biotechnology. Please refer to http://www.responsibleuse.org/resources for the latest guidance available regarding these types of biotech trees.

Sustainable Forestry

There are a number of sustainable forestry management systems that provide basic levels of assurance that sustainable criteria are being met. A list of internationally recognizable systems is available from Metafore: www.metafore.org/index.php?p=Introduction_to_Certification_Programs&s=167.

As noted at the link above, there are a number of systems designed to help manage forests sustainably, including but not limited to:

- The American Tree Farm System (ATFS) is a program for small, private, non-industrial landowners (family forest landowners). ATFS certifies contiguous parcels from 10 - 20,000 acres and was endorsed by PEFC in August of 2008. Online at: www.treefarmsystem.org.
- The Canadian Standards Association is a national standard for sustainable forest management and tracking and labeling certified material. It covers operations in Canada. Online at: www.csa-international.org/product_areas/forest_products_marking/program_documents.
- The Forest Stewardship Council is an international system covering forest management practices and the tracking and labeling of certified products and paper products with recycled content. Online at: www.fscus.org.
- The Programme for the Endorsement of Forest Certification Schemes is a mutual recognition framework for national forest certification standards. Online at: www.pefc.org.
- The Sustainable Forestry Initiative Program is a sustainable forest management standard targeting large industrial operations in Canada and the United States. Online at: www.sfiprogram.org.

Each system is different, with inherent strengths and weaknesses. Note that no particular sustainable forestry management system is endorsed by the Responsible Use: Biotech Tree Principles.

Tools

Templates and worksheets are provided here to help biotech tree users successfully complete actions efficiently and in a robust manner that is capable of being verified if desired. The templates and worksheets that follow are generic and should be accompanied by details specific to the biotech tree use situation. Additional tools are available online at: www.responsibleuse.org/resources.

Example Documentation Template

This assertion document refers to this Practice:

Actions asserted to in this document:

Recommendations asserted to in this document:

1.   Document information
         Responsible Use Principles version: Date:

2.   Primary Responsible Party
         Name:
         Organization:
         Email address: Phone #:

3.   Secondary Responsible Party (required to be In Accordance)
         Name:
         Organization:
         Email address: Phone #:

4.   Relevance to Sections
         Practice:
         Action:
         Recommendation:

5. Proof of Performance
   How performance was achieved (attach additional sheets if necessary):

6. Additional information to fulfill the goal of the practice/s (attach additional sheets if necessary):

Signature of primary responsible party: _____

Signature of secondary responsible party (optional): _____

Public attestation to be In Accordance with the Responsible Use: Biotech Tree Principles: The undersigned is responsible for the fair presentation of the information contained in this document, and for making this information easily accessible to the public. Statements reflect the undersigned's best judgment and are based on the completeness and accuracy of information available and analyzed at the time of completion.

Signature of Responsible Party: _____

In Accordance Declaration

An individual or organization wishing to assert that their use of these Principles constitutes being 'In Accordance' with them can use the following language in communications to stakeholders.

"This report has been prepared In Accordance with the Responsible Use: Biotech Tree Principles. It represents a balanced and reasonable assertion that all applicable practices have been completed according to their actions based on the completeness and accuracy of information available and analyzed on or before (date of completion here)."

Definitions

The definitions printed here are current as of this document's printing date. Updated definitions are available online at: www.responsibleuse.org/resources.

Biotech Tree

The Institute of Forest Biotechnology (IFB) defines a biotech tree as a tree developed through genetic engineering or which contains discretely engineered DNA. This definition is intentionally inclusive of both the process (developed through genetic engineering) and the resulting tree (containing engineered DNA). The IFB considers biotech tree offspring to also be biotech trees unless it can be rigorously proven that such offspring does not contain genetically engineered DNA. If a biotech tree is crossbred with a non-biotech tree then the resulting offspring may or may not contain the engineered genes present in the biotech tree parent, and therefore it is unknown if the resulting tree contains engineered DNA, in which case the IFB would consider this offspring a biotech tree.

Biotech Tree Products

"Anything that is collected or harvested from a biotech tree, or the tree itself." Some examples include, but are not limited to, the following products, if they originate from a biotech tree: seeds, fruit, pine needles, leaves, sap or syrup, tree branches and stems.

Contained / Containment

"A set of controls including the safe methods, equipment, and facilities needed to reduce the potential of uncontrolled interactions of people and the environment with biotech tree materials."

Devitalize

"Rendering biologically inactive." Cease life processes. Complete cell death. Unable to propagate and grow.

Peer Review

"The process of subjecting work, research, or ideas to the scrutiny of others who are experts in the same field."

Gene Flow

"The transfer of genes from one population to another." In the context of this initiative it refers to genes from a biotech tree being transferred to non-biotech trees.

Gene Function

"How a gene and the proteins it encodes behave in the intact organism." In the context of this initiative it refers to genetic changes in biotech trees as compared to non-biotech trees and the resulting difference in the tree itself from that genetic change. This concept is important in understanding the difference between novel gene function and familiar gene function:

- Novel gene function has more risk associated with it because the change in the biotech tree is not completely known at the outset.
- Familiar gene function has less risk associated with it because the change in the biotech tree is somewhat known at the outset.

There is a spectrum of familiarity of gene function that ranges from completely novel, which would be absolutely unknown, to completely familiar, which would be well known.

Greenhouse

"A structure in which temperature and humidity can be controlled for the cultivation or protection of plants." In the context of this initiative it refers to areas where trees can grow in an enclosed space without concern that they could establish or transfer genetic material to the outside environment.

Lab or Laboratory

"A room or building equipped for scientific experimentation or research." In the context of this initiative it refers to an indoor, highly enclosed space where research is accomplished and where genetic material cannot transfer to the outside environment.

Living and Modified Organisms

"Any living organism that possesses a novel combination of genetic material obtained through the use of modern biotechnology" is a term created and defined by the United Nations Cartagena Protocol on Biosafety to the Convention on Biological Diversity. Montreal, 2000. Available online at: http://www.biodiv.org/doc/legal/cartagena-protocol-en.pdf.

Industrial Biotech Tree Products

"A biotech tree that either produces material for industrial applications or accumulates toxic material." In the context of this initiative industrial biotech tree products are novel trees with specific industrial uses. The resulting biotech tree is usually a vector to produce material for industrial applications or to accumulate toxic material. Applications could include environmental remediation of soil contaminated with hazardous waste, and production of pharmaceuticals, unique biofuels, or chemical feedstock intermediaries. This technology is on the very cutting edge of forest biotechnology today. Please refer to www.responsibleuse.org/resources for the latest guidance available regarding these types of biotech trees.

Materially Different or Materially Changed

"Readily distinguishable as significantly different." In the context of this initiative it refers to a biotech tree or its products being changed enough from its non-biotech tree product counterpart so that it warrants a distinction by the user. The level of what is warranted cannot be strictly defined at this point, but it should be based on stakeholder interactions and expert opinions.

Risk

"The probability that an action or event will have an adverse or beneficial effect." In the context of this initiative it means there is a potential for a negative result, but both the result and the degree to which it occurs is unknown.

Spatial

"Occurring, relating to, or requiring space to exist." In the context of this initiative it simply means the extent, or area, of a population of trees.

Stakeholders

"A person or group that has an interest in something." In the context of this initiative, it is anyone who wants to make a positive contribution to the ongoing improvement of the Responsible Use: Biotech Tree Principles. Therefore, you can be a stakeholder. If you would like to beneficially participate, please contact us at www.responsibleuse.org.

Traceability

Refers to the completeness of the information about every step in the biotech tree value chain. It is the ability to verify the history, location, or application of a biotech tree item through documentation in a way that is verifiable.

Committees and Contributors

The Responsible Use: Biotech Tree Principles have been made possible thanks to a number of people and organizational support. The list of individuals that made this work possible is available online at www.responsibleuse.org/process.

IFB's Board of Directors

The Institute of Forest Biotechnology's Board of Directors enthusiastically supported the development of these Principles. Both past and current Board members have played an active role in the Responsible Use Initiative.

Implementation Committee

Includes experts who contributed an extensive amount of time to develop the Practices, Actions, and Recommendations that make up the functional part of these Principles.

Initiative Sponsors

Sponsors have contributed the funding necessary for the Institute of Forest Biotechnology to manage the development of these Principles through its Responsible Use Initiative.

- Biofuels Center of North Carolina
- MWV
- North Carolina Biotechnology Center
- Weyerhaeuser Foundation

Stakeholders

Numerous stakeholders have reviewed various versions of these Principles during their development.

Notes

Notes

www.ingramcontent.com/pod-product-compliance
Lightning Source LLC
Chambersburg PA
CBHW041533280526
45792CB00004B/1487